MW00801072

A VOICE FOR THE PLANET,

A VOICE FOR HUMANITY

BY

ANNA MARIA GALLO

A Voice for the Planet,
A Voice for Humanity
by Anna Maria Gallo

Sunstar Publishing, Ltd.
116 North Court Street
Fairfield, Iowa 52556

Cover Design: Heather Chicoine
Editing: Elizabeth Pasco

Library of Congress Catalog Card Number: 95-071070

ISBN: 1-887472-00-2

Readers interested in obtaining further information on the subject matter of this book are invited to correspond with
The Secretary, Sunstar Publishing, Ltd.
116 North Court Street, Fairfield, Iowa 52556
More Sunstar Books: http://www.newagepage.com

I dedicate this book to

Mysun

Dove Shadow

and Rainbird

for their unfailing patience and wisdom

Contents

AUTHOR'S NOTE

We live at a time when life is very active yet incredibly excessive. We live at a time when we need reminding of the priorities that will sustain us, that will nurture us, that will support life. We live at a time when we must recognize that we are indeed more than a body of flesh— we are a body of spirit as well, and our lives are about the unfolding of this spirituality. We live at a time when life can be very exciting, very miraculous, very powerful for those who decide to engage and live within their spirituality.

There was a time on our land when a rich spirituality existed. It was the time of our brothers and sisters, the native American people. Throughout the book I make reference to the native people, for to me their spirituality contains a profound "purity" of knowledge and a great "intensity" for the value of life. They are held in mind and spirit as a great source in the creating of our own deep spirituality here and now.

It seems our essence unfolds when we direct love and respect toward ourselves and others. *A Voice For The Planet, A Voice For Humanity* was written to motivate and create a sense of courage for those who dare to leap into the discovery of themselves. Each one of us must find our own purity of knowledge, our own intensity for life. It is not out there in the world of form; the mystery lies within. Discover the mystery then pass it on!

My Story

In 1963-64, during my seventh year of catholic school, my religious education was interrupted by a major revelation: "that all of life was a state of mind!" In the moment, at age twelve, when I received this profound message, I realized that there was nothing in life to fear, especially the journey into heaven or hell. The message left such an impact on me that from then on, I knew that some day I would have to communicate it to others.

As time passed I put aside that significant moment and went on with my growth and my catholic education. It wasn't until 1978, some fourteen years later, that I consciously took an interest in mysticism. My journey, which led me to the study of ancient sciences, culminated in many significant spiritual experiences throughout the years. One of the most enjoyable experiences was teaching about personal power, for it triggered a deep feeling within me, allowing me to relive the revelation that I had felt as a twelve-year old.

The desire to nurture, encourage and push humanity into self-awareness has been a strong passion all my life. As I actively exercise this passion, this interlude allows me to see all through the eyes of the God-Force principle and to truly experience the passion of what God is—for this I am forever grateful.

Any questions or comments may be addressed to the following address:

Post Office Box 711
Mount Airy, MD 21771

To everything there is a season,
and a time for every purpose
under heaven.

A time to be born,
A time to die
A time to plant,
A time to reap
A time to kill,
A time to heal
A time to laugh,
A time to weep
A time to build up,
A time to break down
A time to dance,
A time to mourn
A time to cast away stones,
A time to gather stones together
A time of war,
A time of peace
A time of love,
A time of hate
A time you may embrace,
A time to refrain from embracing
A time to gain,
A time to lose
A time to rend,
A time to sew
A time to love,
A time to hate
A time for peace,
I swear it's not too late

CHAPTER I

WELCOME TO YOUR POWER

The Search Within

These words are written for those of you who hunger.
        Hunger for peace.
Peace within yourself, peace with your loved ones,
        peace within the world.

It dwells in all of us but it cannot be described with words
        for it is a feeling—a yearning,
        a desire for love.

Many experience life in a searching pattern usually for
        something or someone.
The feeling to be needed through the act of control
        has been a dear price to pay for claiming an identity,
        when all that is wanted is "to be loved."

Know now that what you are truly seeking,
        is a deeply disguised longing
        to discover and understand **your self**.

The healing process is the "well Being"
        of your body and your mind.
What you eventually become, is your personal potential
        "Realized" within, and about yourself.

The key word here is "Realized,"
        for when one is in a state of realization,
        POWER is unleashed,
        clarity appears—things are made real.

Reality changes.

Think about it.

It's Closer Than You Think

Look around, observe your surroundings carefully.
        Notice, what upsets you?
What makes you feel depressed or confused?
Now look again
        —see what brings on the feelings of joy
        —notice what makes you feel secure and safe inside.

The feelings of happiness are emotions
        that are felt occasionally within us,
        but for most, the empty emotions of loneliness
        are felt too often.
Remember, the feeling of comfort lives *inside*
                and is waiting to be discovered.

Simply opening up to it will unleash its presence,
        as if this feeling inside was an entity
        apart from you.

As a discovery,
        it can be compared to a new born child.
Being the beholder,
        you can be compared to a first-time parent
        giving birth to the miracle of life.

A loving parent embraces their child
        as a self loved person embraces this feeling.

You won't be disappointed.

Value Yourself

We shall call this feeling a "tone"
because it resonates to a particular mode.
This tone is your personal field of energy,
your personal avenue of expression.
Without it we would not exist,
however, without awareness of it,
one exists poorly.

The repetitive action of turning outside oneself for
validation has only contributed one factor
to the human persona,
and that is the power of weakness.

Our outside influences have served us well.

They have been a valuable learning tool.
Unfortunately, in exchange,
we have received an existence in limitation,
and for some,
an unsatisfied—sometimes helpless,
even emotionally and physically damaged life style.

Humans are powerful beautiful beings.

Our endurance for living, without understanding life,
has been an admirable feat indeed.

This power within has been misunderstood
and yet our ignorance of this power
could never negate or lessen its existence.

It has always been with us.
IT IS US.

The message is, and always has been,
to understand it, to use it for our happiness,
and for the purposeful good of all concerned.

Reach for it.

Removing The Blindfold

This power which appears to be subtle in nature
actually is subtle in its application,
yet profound in its results.

Everything around us appears to be molding our lives.
In truth, it is the reverse.
It is the way we feel that is molding our lives.

Feeling camouflages this power and the irony is,
they are one and the same.
We have lived our lives in the reversal of this
magnificent power.

This outcome, unfortunately, has allowed us to become
dead in spirit.
We cannot stop the flow of feeling nor thinking.
One can not stop creation.

All of these elements are connected
causing life to be viewed from a personal perspective.

Living from a perspective of what is seen,
rather than from an "awareness" of what is felt,
can only create more of the same.

Watch your thoughts.

You Are The Creator

Life is a wonderful array of molded thoughts and feelings
which have been allowed to express themselves
through the phenomenon of external appearances.
(Thus creating a world of tangible form and events.)

Form and events are created from feelings and thoughts.

When we hear the word "Power" we usually
associate feelings of achievement or success.
This can be a great misunderstanding,
because one never becomes fully satisfied.

The fulfillment for experiencing power does not exist
in the achievement of material possessions.
It exists in the understanding of how we create
what we possess.

Again, fulfillment does not exist in the identity
of what we become.
It exits in the "power" that we can become it.

This attitude allows a true sense of responsibility
to take shape.

When taking the responsibility for what has been created,
life allows us the grace to step away from our creation,
view it ... change it,
and by choice decide whether we want
something different or not.

Be flexible.

Power In Reverse

When we do not take responsibility for our creations
negative hurtful situations are formed
from a lack of understanding and
directed toward others.

These situations are kept alive through
anger—resentment—fear—guilt.

As these emotions continue to run rampant,
fueling themselves with their own feeling,
our "power" begins to work against us.
Its compounding energy produces more painful events.

This habitual emotional strain can only result in
the dissolution of the physical body.

Get off the roller coaster.

Expanding The View

The material expression of failure and success has
been dreamt of,
experienced by,
and available to all.

Life has more mystery to be explored than the goals set
for failure or success toward a given direction.

Living the dream to succeed or fail is an important step
to embrace in our journey.
Getting stuck in this small focal point of concentration,
however, can prevent us from experiencing
the true gifts of life.

Our journey needs the ingredients of
growing and *sharing* at the same time.

Our life's essence is MOVEMENT.
Our ticket is PASSION.

Emotion—the inward movement of our passion—
is the power behind creating reality.
If we move inwardly through feeling
it will show itself outwardly in form.

To spend our time in repetitive situations is to die
S L O W L Y.

To live for the purpose of survival causes us stress
and strips away our soul.

To live in boredom (the signal for change),
is to deny ourself our power.

The feeling tone within is waiting to produce itself
into new expressions of life.

Break from the fear and use courage to unleash
the talents hidden within.

Take more risks.

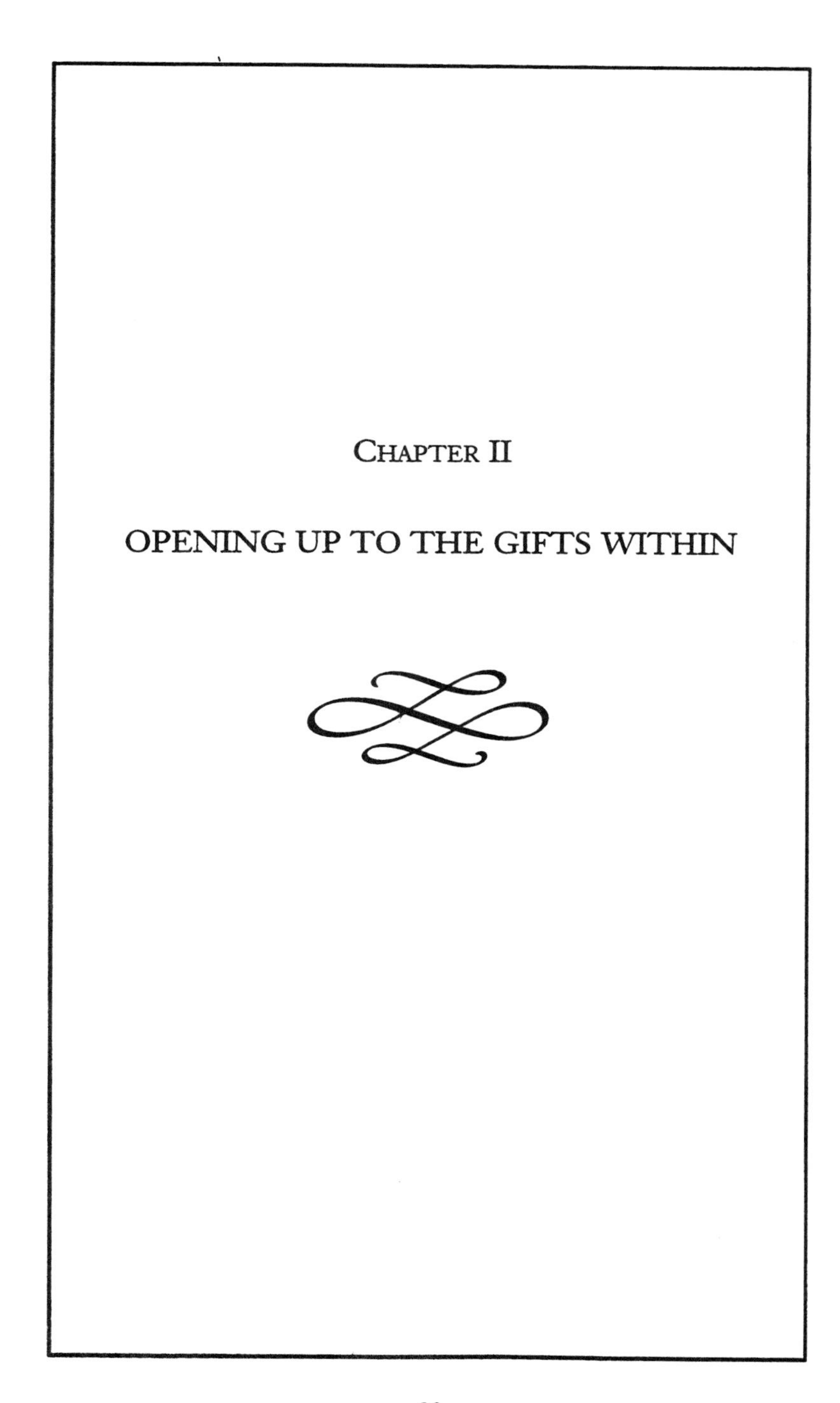

CHAPTER II

OPENING UP TO THE GIFTS WITHIN

Vision Versus Materialism

For a long period we have used our power unconsciously.

Unfortunately this has led to a lust for
a lifestyle of materialism.

"Unconsciousness,"—to be unaware—
does not comprehend inner power.
It comprehends outer power.
(What you see is what you get.)

Consequently we have created a world of greed
and illusive beauty, making our power destructive.

What keeps us spinning our wheels, so to speak,
and bonded to this mode of expression?
—our distrust in the power of our own intuition,
—our denial of a life of self worth,
and most importantly
—our refusal to take personal responsibility
for our actions.

For the most part, on a individual level,
this limited expression of life, as we know it,
is experienced through emotional and physical pain.

Our prolonged use of negative thinking
along with the constant "belief" in limited reality
leads us to mental suppression.
Mental suppression causes mental depression
and stagnation.

Depending on the strength of an individual's will,
this attitude could compound into manic depression.

In severe cases it has been known to cause
mental and or physical suicide.

Fortunately, the human spirit was not designed
to live this way,
yet our culture accepts this conduct
as a "natural" function of society!

Just think, if we were to open to our power,
we could do away with therapy as we know it
and create a new world of communication.

Dare to try.

Emotions That Bind Us

Transforming within requires looking
at some of the emotional roles that hinder us
from tuning into this feeling of love and safety.

Let us begin with the emotion of anger.
Anger is usually expressed to someone or something
when our integrity is on the line.

Often anger develops over a period of time
because opinions were sacrificed instead of expressed
in the moment of a situation.
If our opinions are not voiced the moment they arise,
they become suppressed and anger swells up within.

Anger is usually held back for fear of hurting someone
or being hurt by someone.

We can gracefully express anger by first separating our anger
from our true feelings about the individual.
What we are angry with is the action
or choice of the individual,
not the individuals themselves.

Anger can be communicated simply
by directing our feelings out loud as they are being felt
and then moving on.

When an outburst of anger occurs it is in truth
a release of personal suppressed emotion.
It is literally an act in spiritual survival for attaining
and reclaiming our own power.

Unfortunately we judge our anger by feeling guilty.

This is because we don't honor our feelings
from moment to moment.
The set back here is the misconception that anger is
an attack on others.

In truth
anger must not be confused as a lack of love.
It is an act of love,
for one is often feeling love and anger
at the same time.

Once our sense of power returns
anger dies out much like a burning fire
... slowly but surely.

Next—the inner turmoil of victimization.

Victimization is the lack of inwardly knowing
our ability to accomplish our desires, goals,
or expectations.

Victimization places us in a position of dependency.

To reinforce inward knowing, we must engage
the courage to follow through on our ideas.

The choice to act, rather than judge
or rationalize our ideas away
creates great inner power.
The end result is the ability to accomplish our desires.

When we choose dependency we are usually victims
of another's reality, and in time lose our own life.

Dependency is caused by fear and self-imposed
limitation to survive life rather than live it.

Find the courage to take risks,
for the promise of spirit is to take care of you.

The emotion of guilt is a crippling one and indeed,
probably the most detrimental of all.

Guilt is often chosen in a time of weakness
yet it continues to give away our power.

Guilt, on a feeling level,
is simply a conscious reminder
not to repeat the same attitude.

Usually we associate guilt with our lack of cooperation
in some way for or toward others.
Judging our actions not to cooperate causes the guilt.

Guilt leads to unhappy, undeserving choices
for ourselves, causing painful regret.

To eliminate guilt, we need to change
our approach and perception
to one of honoring any choice made
as the best choice at the time.

This teaches us the value of unconditional love
for we are accepting the behavior
of whoever is involved.

What makes guilt so dangerous is that
it has a much deeper magnitude.
Guilt can seep into everything,
not just our reaction toward people.

If allowed to control our lives
it will forfeit our happiness,

stop our energy flow,
and create emotional states of unworthiness.

End the guilt from consuming life,
all it takes is a shift in attitude!

The largest culprit of all,
fear,
resides in many negative and destructive emotions—
anger, victimization, guilt, hate, resentment, hostility,
shame—an unending list.

There are also many facets of fear;
fear can limit life because it stops action.
It can destroy life because it stops joy.

On the other hand, fear can motivate
and become a stepping stone for courage and change.

An important thing to remember is that
fear always afflicts when we live in the ignorance
of knowledge or self awareness,
yet it becomes a tool when faced or challenged.

Eliminate fear by keeping an open mind
and a desire for new experience.

Let go of the resistance.

You Are Free

It is of great importance to realize
that all of these negative emotions
are generated from a lack of personal expression
and from our "personal belief systems."

Our systems of belief are extremely powerful,
they control the direction of life.

The beliefs we live by are based and formed
on trusted information handed down by many people
coming in and out of our lives.

At an early age this can be detrimental
because of our vulnerability.
Later unfortunately, we accept too freely
this information and relinquish our choice
to experience it objectively.

In time, through habit,
we become enslaved to our belief systems
and unaware of our heart's desire for change.

It is considered most wise to take time off from belief systems,
for they cannot be replaced
unless we make ourselves aware of them.

Once we are aware of our beliefs we can proceed
to replace them through the process of testing them.
Remember, the discovery of this feeling
is of utmost importance for it contains
the essence of who we truly are.
It holds the key to pure creativity.

It is activated by directing our attention inward and

utilizing the many insights available to us.

These insights within us are magnificent pearls of wisdom.

When we choose to use them consciously,
the very act itself serves as an act of love
allowing the insights to become gifts.

Only through experiencing can we begin to understand fully
the depth and beauty spoken in the language of spirit.

Learn who you are.

The Gift of Intuition

Intuition or "inward knowing" is
recognizing
and *responding to*
all information that is received inwardly
without judgment or analysis.

It has been known as our "gut feeling."

It functions at its best when we are in a state of
relaxation.

When we are calm,
we automatically disempower ourselves
from our current state of reality
—we are *flowing* through time.

This calmness creates the ideal condition
for intuition to occur.
When we are serene,
we are in a state of receiving messages and insights
through feelings.

Intuitive messages usually take the course of giving us
information based on personal activity.
Therefore
they involve messages
concerning persons, places and things.

Our intuition serves us best when used for direction,
but the direction must be followed through
or the intuition will weaken.

Intuition has also been known to break through
while under great stress as guidance.

However,
it requires a great intention for truth
to draw it out.

Overall, being composed within
has many benefits
and is the best path to follow.

Go with the flow.

Exercising Affirmation

An affirmation is the act of
propelling thoughts into existence.

We are always in the state of affirming something,
whether negatively or positively.
The significance or intensity we place on thought can
either help us
or hinder us.

Turning a negative emotion into a positive emotion
requires monitoring our thoughts.

"Checking into" our self
or becoming aware
is our goal.
Taking the time
daily
to know how we are feeling will be very productive.

Our reality is created by our emotional thought forms.
When we begin to play with happy thinking
we will recreate
and change our lives.

Affirming our life in positive ways
can lead to a healthy attitude
and the restructuring of our lives.

When an affirmation is going against an ingrained belief
about oneself,
the affirmation will not work.

What is needed in this situation is
a desire for change,

and a *surrendering* to an attitude
implying that change is indeed possible.

In this type of a rigid personality,
written affirmations are encouraged.

Practiced affirmations are most beneficial
when we feel the need to do so,
they almost never work when forced.

In time they will become instinctive in behavior for
we will begin to "believe"
what we are affirming.
This process is very effective because
we always accomplish our needs
when acting from a space of love.

The gift of affirmation gives us the power
to "know" who we are,
and in that knowing
we become *safe*
to live our lives in joy.

Believe in yourself.

The Exercise of Momentary Living

Directing our attention to the present moment
allows us to step out of the past and out of the future
with our mind.

When we are not in the moment our mind wonders;
this usually makes us feel the need to
"figure things out."

In this "control" issue it is very clear that
the "safety" principle is struggling,
and in that struggle it is creating
manifestations of unwanted
probabilities and possibilities
of an undesired nature.

There is an affluent amount of events happening in
moments that are remarkably precious
and, sadly, are usually missed.

These priceless moments are the splendor of life;
when not experienced
they are lost forever.

Not living in the moment leaves us meandering through life,
questioning indeed, "Is this all there is?"

To live in the moment
we must express our feelings.

When we don't live in the moment
we deny ourselves growth.

A good exercise for relinquishing the "automatic pilot"
and getting in touch with this "mode" of living

is when driving in a car,
because driving forces us to stay present.

While driving, besides using the physical sense of sight,
we can experiment with smell and sound
to focus our attention outside the vehicle.

Interestingly, after reaching our destination,
we might find that time has been altered,
and our state of awareness changed.
We may feel less stressful and quite possibly
amused.

Some of the qualities that will develop within
are patience,
wisdom
and compassion.

Living in the present moment becomes a great teacher,
but most importantly,
it eliminates the presence of fear!

SEE what you are missing.

The Gift of Meditation

Meditation when practiced is a personal enlightening
experience
—a great tool for total relaxation and healing.

When we consciously meditate we shut out
the outside world of material illusion
and keep it from our direct attention.

The result allows us to explore and play
in a completely different world than the one
we are "physically" accustomed to.

Meditation "loosens" the validity of form.

It challenges the reality of the physical world as
being the only world,
and reminds us of how unlimited we really are.

Many techniques have been created for meditation,
some more complex than others.

We only need to choose a technique that is simply
in harmony or alignment with our feelings.

To get started,
being still for fifteen to thirty minutes daily
will give the best results.

Concentrating on our breath
will clear the mind of mundane thoughts.

If there is strain,
gentle music may allow all thoughts
to surface in freedom.

Prayer is a form of meditation
but not to be confused with
"free flow" meditation.

Meditation is a releasing.
Its purpose is to create a sense of well being,
while strengthening the voice within.

In time, it will become second nature
and the connection will be very *personal.*

The gifts available through meditation are endless:
they are abstract in emotional form
as well as visual form.

There is beauty within.

The Powerful Gift of Dreaming

Sleep dreaming serves us in two vital ways;
by providing emotional release
and by exercising the use of manifestation.

Our pain and fear accumulates
through the stubbornness of not letting go
and through the stubbornness of
not forgiving ourselves and others.

Our dream states can act as a dumping ground
for our painful and fearful emotions.
Be thankful for nightmares!

When we are not releasing in our dreams
we are creating and traveling.

Living in a contented fashion allows our dream state
to act as a catalog,
where we can do our window shopping, so to speak.

We find great maneuverability in our dream world.
Traveling to other realms of experience
is a common occurrence and very much practiced,
along with visiting other souls
in this world and others.

Not remembering our dreams can be caused
by many personal and perhaps complicated reasons.
Knowing that our will is the strongest force in the universe,
we might give ourselves the suggestion
"to remember" before sleeping.
Eventually this will result in remembering the dream.

Investigating our dreams can cause profound therapeutic

results leading to wholesome individuals.

Since most of our dreams are communicated
through the language of symbols,
they have a tendency to evaporate rather quickly,
therefore it is wise to keep a journal of our dreams
for future reference.

Besides night dreaming, daydreaming is also greatly involved
with our ability to change the exterior world.

Daydreaming is a form of meditation
that should be encouraged, not discouraged.
When we daydream we are usually in a state of bliss
and genuinely aligned with our innermost desires.

To be in a fantasy is the same as day dreaming.
To turn inward is to use the imagination,"image in."

Using our imagination expands our futuristic vision,
for when turning inward, we become
inspired,
inventive
and creative.

Let your thoughts
pertaining to your heart's desire
flow.

Dream, Dream, Dream.

The Force of Nature

The most powerful and yet graceful in its design
is the great provider
NATURE.

Nature is the only catalyst outside ourself that can prompt
this feeling of love and safety
into conscious awareness.

Nature is conscious,
she speaks and listens.
She converses through feeling.
She contains the passion of freedom,
the sensations of wonderment,
the genetics of wisdom,
the compassion of love,
and reflects them all back
to YOU.

She reveals her creativity through
her ever changing seasons,
and her love
by sharing her very life to extend all life.

Nature communicates the message that
we are connected to her.

Observing mother nature stirs this
sleeping power within us.
It allows us to perceive what is important.

Nature is *power* ... in motion.

Embrace her through feeling and allow this power
to grow within us.

So penetrating is this power that it will
spontaneously STOP
all feelings of stagnation.

Nature's power arouses emotions of courage within,
motivating us to eliminate unwanted barriers
in our life.

So active is this power
that it will allow thoughts within
to expand into concepts
that have no words.

So comforting is this power that it emotionally
nourishes us in times of need.

So loving is this power that it requires no demands
from anything to prove its existence.

This power is the God force energy that lives
and has its being in all of us.

It is a source driven by feeling
penetrating all life,
all matter.

This source is "absolute" in power and holds all life
seen and unseen together.
Its journey is a gift given to us
and it is to be fully experienced.

Walk through your kingdom.

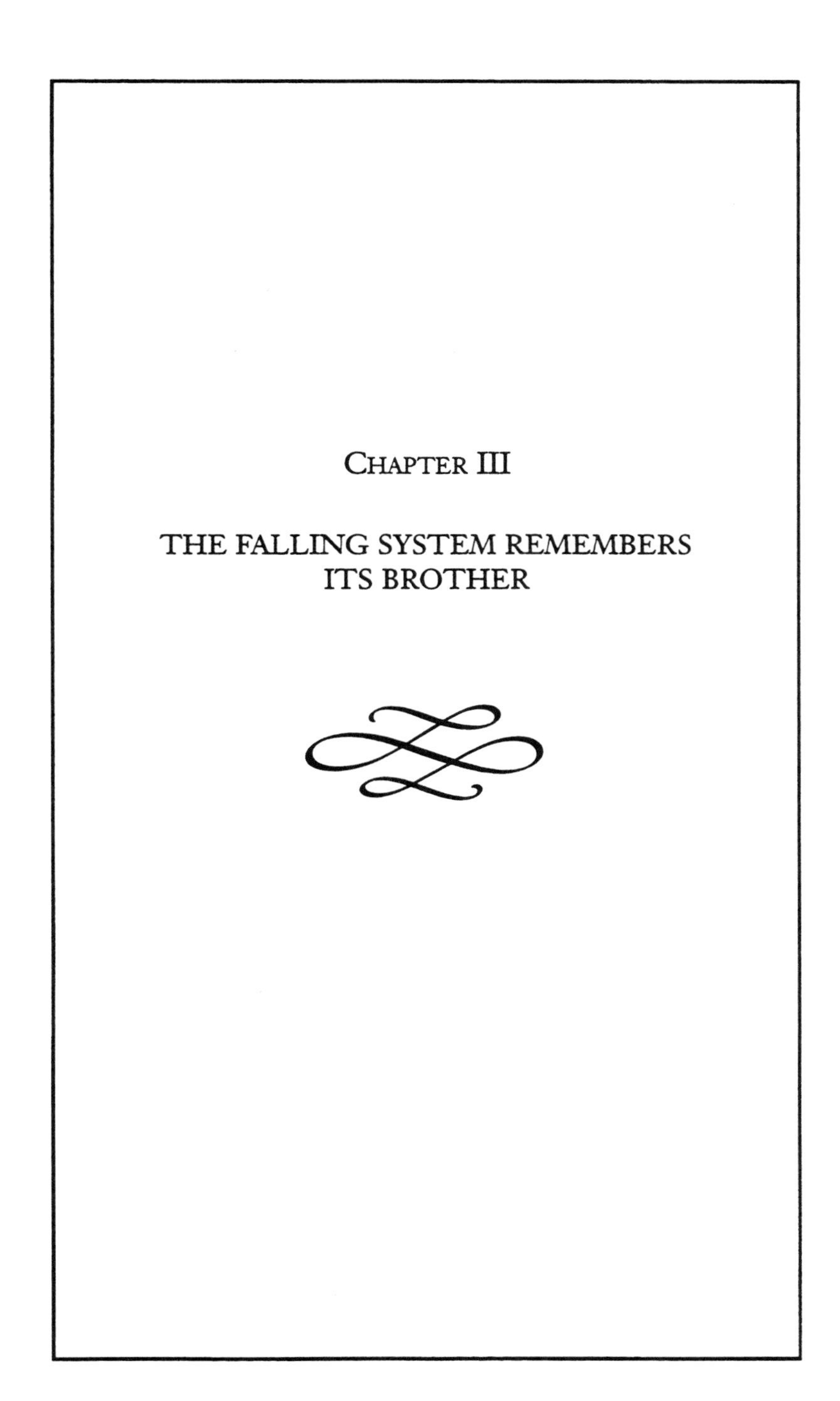

Chapter III

THE FALLING SYSTEM REMEMBERS ITS BROTHER

We Are One

Contemplate the concept—
WE ARE ALL A PORTION OF A WHOLE,
INDIVIDUALLY FOCUSED.

There is an irrefutable law in effect that can not be altered
that is, each one of us is important,
holds the same power,
and is connected to each other.

We are connected physically, emotionally, and mentally.
What we think, feel, and verbalize, alters all life
seen and unseen!

It only takes one person, one at a time
to make great changes.

As we begin to open our mental capacities to
receive spirit in our hearts and mind,
we literally invite others to do the same.

Our growing processes primarily and most strongly affect
the members of our biological family.

They, without effort, begin to receive the growth
that was accomplished by others simply through the
willingness to open and accept new concepts of
thought coming their way.

These new concepts are transmitted through the process
of personal revelation,
however they must be embraced and applied
for changes to occur.

Receiving the concepts is not enough;

they must be explored.

This connection that we all share allows us to
participate in our own evolution,
and if we choose to,
with awareness.

Notice how the word LOVE
is included in the word
evolution,
stressing the importance that
love is the path for survival
for ourselves and our planet.

Choose to love.

Life Is Love

When love is taught to be repressed,
life creates devastation...
the flow of life is to live with love.

For the earth to restore herself, she needs help from
the consciousness that abused her.

There must be an end to the dysfunctional consciousness
on a massive level.

There was a race that knew the importance of nature.
Reflect back and remember...
the American Indian.

As a race they were humble in spirit.
Their rapport with nature clearly demonstrated
reverence and love for all life.

Through their openness to receive from nature,
they in turn were gifted with her inspiration.
This inspiration provided them with a world
rich in harmony, both physical and spiritual.

This interaction sustained the earth,
for it was a life style of giving and of taking.
Balance was created,
life became a "natural" state of affairs,
"a love affair,"
of life giving life.

Respect the provider.

Nature As Our Teacher

When we are in alignment, in harmony with life,
we are understanding
and living in cooperation with and for nature.

The native American Indian was "aware"
of his relationship with nature
and demonstrated it in what was called his "medicine."

Their medicine was a healing power derived from
inward *knowing*.

Medicine was the cultivation of their intuitive knowledge.
The power used, was source, or spirit,
the knowledge received was intuitive communication
between the source and "vehicle"—man.

The Indian man received his wisdom proudly,
the tool used was nature,
the gifts granted were animal, plant and water.

For the Indian, nothing was ever wasted.

He understood that to upset balance
was to jeopardize himself.
He started his day
acknowledging and praising the sun
for its service in sustaining life.

He ended his day acknowledging his heritage to the stars,
honoring them for their wisdom
in sculpting his future.

He lived in "total" gratitude of life,
accepting all life as a gift.

Every moment was reverenced in the feeling of spirit,
from his hunting to his celebrating, even to his dying.

The Indians were always aware of their connection
with the earth
and evolved "with" her, not against her.
The physical makeup of our bodies holds the properties
of the earth within.

America's great healer, Edgar Cayce, was quoted saying
(in his reading #2396-2),
"There is within the grasp of man all that in nature
that is ... an antidote for every poison,
for every ill in the individual experience
if there will but be applied nature, natural sources."

Our earth has been violated by a consciousness that intruded
upon the American Indian long ago.

This consciousness lacks humility,
lacks insight.
It continues to feed itself through greed.

The American Indians and all Indians throughout time
symbolically represent the life force.
Most importantly, they are a reminder to us,
that they were and we are the guardians of this planet.

Their legacy needs to be recaptured
and incorporated into our way of existing
for the survival of ourselves and our children.

Let nature show the way.

Man-Made Disasters

Greed ... an act of ignorance
displaying a lack in the understanding of our
spiritual power.

In the name of greed we have created a global rise
in temperatures this century.
We are emitting poisons into the atmosphere
at a rate that will change the climate
more rapidly than it would have changed naturally
during the last 100,000 years.

The industrial revolution
is considered the most important contributor
of increased levels of CO^2 in the atmosphere.
When power stations, factories, homes and cars
send water vapor and carbon dioxide into the air
as a byproduct of generating energy, CO^2 increases.

Burning any type of fossil fuel makes carbon dioxide
and is considered the prime suspect
in the greenhouse effect.

Eighteen billion tons of CO^2
are emitted into the atmosphere *each year*.
And these levels are expected to *double* within
the next fifty years!

Other powerful man-made contributors
to the greenhouse effect are
chloral florocarbons (spray Propellants),
also used in foam products and nitrogen oxides,
a byproduct of hot combustion as in car engines.

It is estimated that 40% of CO^2 is being absorbed

by our ocean waters
and this absorption can go as high as 85%.

The interactions between air, water and temperature
influence climate conditions,
and our climate system is a complicated mess.

The implications of a warming planet threaten
both human lives and natural resources.

Under research and experimentation,
crops grow faster with less nutrition
in larger amounts of CO^2,
while weeds competing with crops get stronger
each decade increasing pest damage
and upsetting agriculture as we know it.

Let's not exclude tropical deforestation
where 27 million acres of trees are lost *each year*.
Instead of planting trees to absorb CO^2,
we are destroying world forests.

The full extent of damage is unknown,
unknown.

What of our respect and responsibility for our animals?
They are our greatest treasure.

Our animals and insects maintain the delicate balance
of nature through there instinct to survive.
How and where will they direct their instincts
in a mutilated environment?

How will we direct *our* lives
after we have eliminated them?

Greed

Greed is BIG business with costly mistakes.

On March 24, 1989 the worst environmental disaster
in the nation's history shattered our complacency.
The Exxon Valdez, an oil tanker, spilled
11 million gallons of oil in the Alaskan waters.
The spill covered one thousand miles of shore line,
killing one half million birds
and thousands of mammals in all.

What could have been a controlled oil spill
turned into a "political coverup."
Precious energy was used to protect an "image"
rather than to clean up a disaster.
This horror story is but one of several spills
that have occurred over the years.

Today we carry seventy times as much oil on our tankers,
and the damage of these spills
is expected to go well into the next century.

Sadly, we have tolerated far too many disasters.
Why aren't we seeing the warnings?

Our mammals are communicating to us
that our waters are polluted
by beaching themselves on the shore lines.

Our atmosphere is communicating to us
by its erratic weather patterns.

We need not look far
to see our destructive behavior communicating to us,
it is always in view through our media.

We hear of radiation poisoning in the air,
bacteria contaminating our waterways,
more oil spills, viruses increasing,
families violent against each other,
suicides being ignored,
children killing children.

This *must* end.

To have forcefully invaded, disabled and degraded
the American Indian was an atrocity.

To damage and deplete the earth
with her inhabitants is an atrocity.

This *can* end.

All of us hold emotional pain inside.
Let's make a conscious effort not to ignore,
but face and feel our pain.

We must teach ourselves,
only then...
can we teach others.

Now is the time to change.

Spell Bound

Using earth as a reference, all evolves.
"Natural form," life, changes.

The universe and all its participants are included as well.
All heavenly bodies evolve, change,
and effect each other in the process.

The ambiance of the universe is changing
and will effect all living things on our planet.

Life is not just experienced through the physical,
but through the mental and emotional as well.
As a result, all three faculties will be affected.

This phenomenon brings to mind two factors,
one of *intensity*, and one of *time*.

Whatever the experience
—emotional, physical, or material—
everything is occurring with great intensity,
and with an acceleration of time.

Time is running out for all who continue to live
in the attitude of greed and fear.

The attitude of greed carries emotions such as
excessiveness, self indulgence, jealousy,
manipulation, vengeance, extravagance, insecurity.

Because many who feel these needs live under
this enslavement labeled greed.

The attitude of fear carries emotions such as
anxiety, worry, dread, confusion, distrust,

suspicion, misgiving, disbelief.

Because many who feel these emotions live under
the clutches of fear.

Fear breeds greed breeds fear.

This attitude of complacency and negligence
has severed us from our inner spirit ... our integrity
and has caused heartbreaking effects
on our beautiful planet.

Humanity is being governed by an attitude
seeded on greed,
and controlled through the power of sexuality.

This way of life has placed us
totally out of balance with our environment.

Our industrial world system has created a monstrosity
for our hungry, homeless and masses that are
aging alone.

Where is our allegiance to them....?
Where is our respect for the minds being left behind?
Why have we allowed ourselves to feel so helpless?

All life was created through the act of love.
All life is sustained through the act of love.
All life is allowed to exist through the
act of love.

Sow love, reap joy and abundance.

Nature On The Move

To be alive
—to witness our earth evolving
is a grand event indeed.

Our earth is patching up her bruises and moving on.
Interestingly, she is not particularly disturbed
by who is standing in her way.

Nature does not regret anything.

Unlike modern man, she is a *forward* thrusting movement.

Many of us give little thought
about how to survive an earthquake or a hurricane,
or any natural disaster.
We take our physical safety for granted.

Our passion to learn and to become sovereign
has reversed into a slothful attitude.

When a natural disaster occurs,
we expect those in power to correct it.
Unfortunately, those in power
seem to have us right where they want us.

We have fallen behind in the development
of our "internal" power
meanwhile nature has moved on ahead.

During these times we need to understand
our spiritual power.

As humanity stands on the brink of a new story
—a new paradigm—a new existence,

our spirituality will see us through.

Our earth is making
transitional and transformational changes
in structure and harmonic motion.

It is not *the* end,
but an end to old out dated mass belief systems.

Only through knowledge and personal power
can we participate.

Ride on the winds of change.

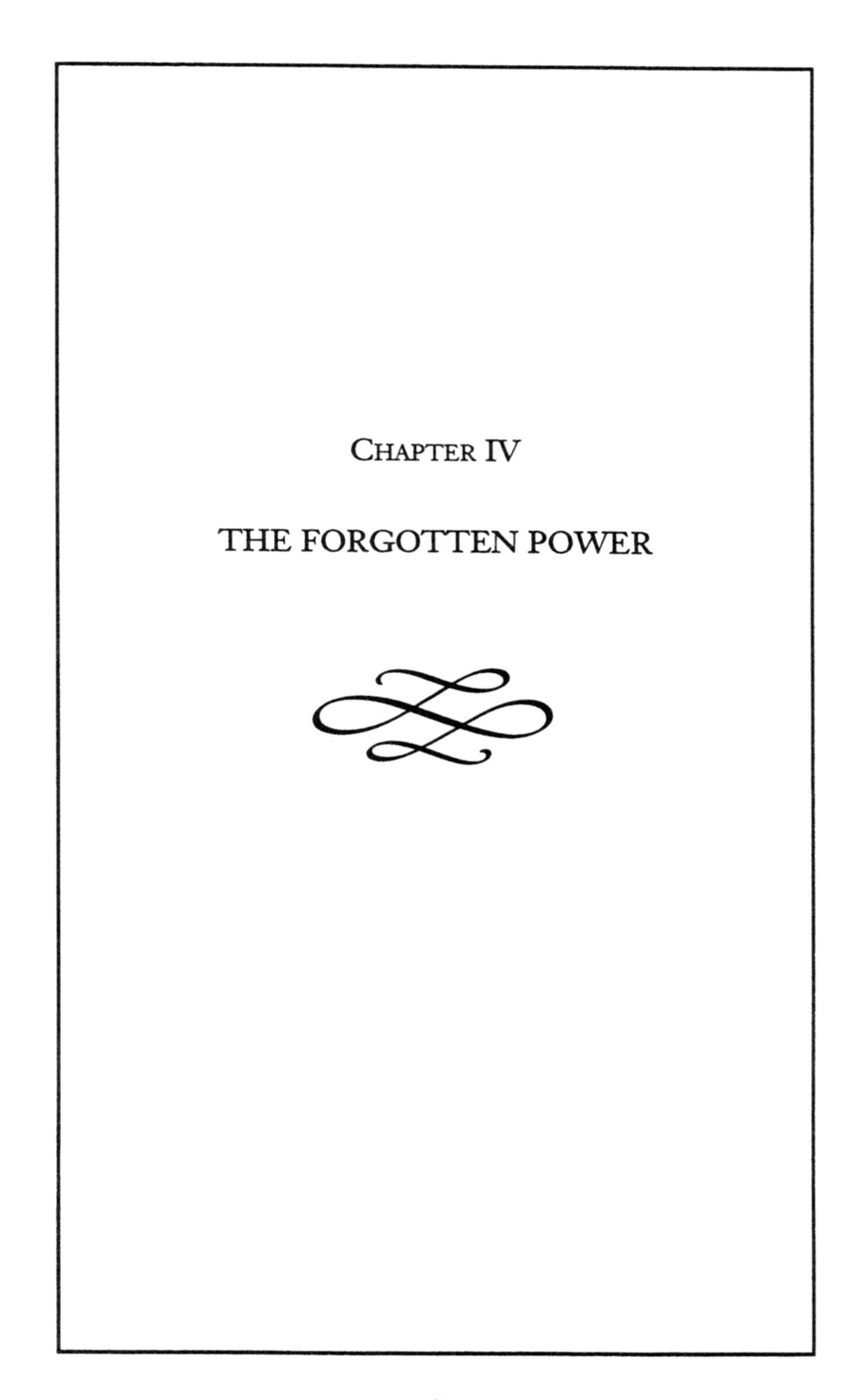

Chapter IV

THE FORGOTTEN POWER

There Is Choice

Those choosing to continue living
in the life style of fear and greed,
will find great difficulty.

Trying to sustain such an existence can only create more stress
because through evolutionary cycles,
the earth is experiencing an increase in energy flow,
causing life on the planet
to amplify and intensify as we know it.

Struggle after struggle will be experienced,
until the need for change is recognized,
or simply "giving up" will be chosen.

Resistance is futile,
only because it works against any condition.
All conditions are created by our own power.
Whether through ignorance or deliberate intent,
they must be lived.

Choosing to flow
with comforting or discomforting conditions
proves to have better results
than resisting them, for they will consume you.

Free will is a gift, use it wisely.

Spirit Knocking At Our Door

Within us we have an essence,
a spiritual presence,
longing to participate and expand itself.

The truest part of ourselves wants *expression.*

If we *truthfully* gazed into the mirror
in search of an identity,
we would have to admit
that we are tired of the games being played.

All games have opposites.
They are merely diversions to the inner essence.
Deep inside, we know that we deserve joy!

Whatever the game,
there is always a counterpart game
waiting to be experienced.

The way to play the opposite game,
is to imagine you already have it.

"Believing" in whatever lackful game you are in,
holds it there.

It may appear to feel awkward,
perhaps even difficult to presume or validate
something to exist, that does not already exist,
but this is how it is accomplished.

Cause produces effect.
The existing game was created
through this process of "embracing,"
and can only be eliminated the same way.

Expectation creates circumstances!
This means looking for,
    acknowledging,
    and placing attention on,
    what makes you feel good.

It takes courage and trust
    to rise above survival games.
It takes awareness not to be influenced by others,
    but its reward is sweet,
    its reward is life!

Give personality the day off, let spirit in.

The Fiber of The Universe

There are laws in our universe
that are electromagnetic in nature.

The word "electromagnetism" means,
"the phenomena associated with the relations
between electric current and magnetism."

There are laws in the universe pertaining to,
and in a "state of order."
These laws are in accordance with
vibrational frequencies.

When we think of the word "vibration,"
we usually associate it with sound.

Also, when we think of "electromagnetic,"
we relate it to electricity used for power.

Well, it may interest you greatly to know that
vibrations and electromagnetic forces
are components of our universe.

All forms of material in our universe,
including us, are electromagnetic in nature.

Everything consists of atoms
and atoms consist of protons,
neutrons,
and electrons.

A proton is an elementary particle
—a fundamental constituent of all atomic nuclei
having a *positive charge*
in magnitude to that of the electron.

The electron is the *negative charge*
and exists outside of the nucleus of an atom.

The emphasis here is placed on the awareness that
we are like batteries.

There is more to us than meets the eye.

Flowing Energy

When we use up our own energy,
we deplete ourselves through stress.

The need to be "recharged"
through proper physical care is extremely important.

Many of us do not take the time to do so,
which leads to a deteriorating life.

When energy passes through us from spirit or source,
our energy flows more rhythmically;
less maintenance is required.

Take a moment and apply this information to feelings.

Referring to the body's nervous system,
a nerve is defined as a bundle of fibers
forming part of a system
that conveys impulses of *sensation and motion*,
between the brain or spinal cord
and other parts of the body.

A feeling is defined as the function or
the *power of perceiving by touch*,
and further defined,
feelings hold the capacity for emotion.

Therein lies the connection
that both nerves and feelings
are physical and electromagnetic!

When we entertain thought,
our thoughts are usually accompanied by feeling.

Thoughts come randomly,
deliberately,
unwantingly,
and forcefully.

Ask yourself what type of feelings
would accompany such thoughts.

If the thoughts are haphazard, unwanted, or forceful,
the feelings most likely will be
feelings of unpleasantness or pain.

If one is basically "feeling" unhappy,
then one is generally giving off unhappy thoughts,
usually "unconsciously."

Feelings send out signals.
These signals are "electric,"
and draw to themselves like signals of attraction.

So, as in batteries, our current vibration
is our current feeling resonating with
an emotional frequency of either
"happy," the positive charge,
or "sad," the negative charge.

Energy is the fuel we use to *expressively* live.

The act of resistance breaks down energy
by causing the energy to lower its frequency.

Resistance always manifests itself
in form as a "negative" situation.

To resonate with a negative attitude,
is to be in the *absence* of the positive charge.

To resonate with the positive
is to be in the *absence* of the negative charge.

When we "consciously" participate with energy,
we have the option to experience
both the negative and positive charge.

Becoming stuck in the negative
can lead to destructive behavior.

Forcing the positive is a misconception also.

A self induced or forced positive behavior
eliminates the negative charge,
and this severs communication between us and spirit.

We need the negative feelings
to serve as a warning or an option for choice.

A starting point is needed to know what is not wanted,
then the energy *shifts*
toward the positive to receive what is wanted.

When we are just flowing pure energy,
we feel the entry of both
the negative and positive charge simultaneously
and in a moment we recognize
what is best for our well being.

This is being in a state of balance,
becoming "whole,"
for *spirit* is present.

This energy of vibrational frequency
exists everywhere in our universe.
It exists in its own creative order

by degree, or scale.

On our planet, the degree or scale
reflects itself by the intensity of feeling.

Since no two people are alike,
neither are their feelings,
making this a very complex world.

Rather than spending time
in fear or victimization,
we have a responsibility, to our world,
to our universe,
to live in joy.

Remember, we are connected with all things.

Chapter V

THE LAWS THAT GOVERN US

The Law of Magnetism

Besides physical laws in our universe,
there are "rhythmic" laws as well.

The one primarily needed to be understood is
the law of magnetism.

We have all heard the familiar sayings,
"like begets like"
"birds of a feather flock together"
"like attracts like."

Well, all these sayings
pertain to the law of magnetism.

The law of magnetism challenges our choice.
Our choice seems to be present,
but when we live in a state of "reaction,"
our choice was not properly examined.

We emanate "vibrational signals" called feelings.
These feelings have a rhythm, or a "pulse."
Rhythm, being a wave of "motion," ... travels.

An example of this might be
recalling an incident while shopping or in a crowd,
and "not" feeling pleased,
or that of feeling "very" pleased,
while standing within three feet of an individual.

An experience of comfort or discomfort is felt,
just by being exposed
to an individual's electromagnetic field.

This is a good exercise for familiarizing ourselves

with the "awareness" of vibration in general.

This law of *rhythm* seeks identity, and identity means,
"the state or fact of being the same;
exact likeness in nature or *qualities*."

"Like attracts like."
So, this law creates an experience
that validates our feelings *through* an event.

The rhythmic law of magnetism is exact in its nature.

If we are feeling bad, or irritated,
something will always occur
to "match" the intensity of the feeling.

Through the process of "thinking" or
"projecting" the irritated feeling,
the irritated feelings can only *mount*,
causing distress on a daily basis.

Using the analogy of a camera,
the *eyes* can be the lens through the *mind*.
The mind housing the image or the story then
projects itself through *emotion*,
onto the screen of time and space,
and by law seeks identity.

The finished product is always
a *direct* manifestation derived from feeling.

What we allow ourselves to be
influenced by or conditioned by,
can be harmful, or wonderful.

Our world is like a mirror reflecting back to us

everything we feel inside.

What occurs in our lives
will always prove and validate
how we express this divine energy.

When our world has become less than desirable,
we must take a modest look at ourselves,
by *privately* acknowledging our weaknesses,
then *committing* to changing them.

For this law to become a natural focus,
it simply requires a *developed* sense of awareness
of what is going on around us.

The way has been made simple, use it and become strong.

The Proper Use of Free Will

What happens in our lives cannot be changed
through *physical action*,
but only through **conscious intent**!

Taking physical action is a short term plan,
and, it almost always back fires!

Choosing physical action to correct a situation
can only take care of the symptom,
not the CAUSE.

Taking care of symptoms
can only result in other symptoms taking over.

Feeling is what created our dilemma
and feeling is what will change it.
Unless we remove the unwanted feelings,
the experience can only reoccur.

The technique of analyzing upsetting feelings
is excellent healing work,
and very much encouraged.

Returning to the past to correct the present
is very delicate work.
It requires a keen understanding of timing,
and a discernment of knowing when to "stop" analyzing.

Returning to the past
can also become an entrapping experience.
It allows repetitive feelings
to remain alive while in the processing;
that is why timing is so important.

Make peace with the past,
and work from a standpoint of "feeling good."

Thinking and feeling about what is wanted,
what is more productive,
will alter the experience much faster.

Free Will is a terrible thing to waste.

The Art of Monitoring Our Feelings

Giving of ourselves to the outside world,
while focusing on personal feelings,
can be a tricky task indeed,
but very beneficial.

The act of monitoring our feelings requires us to live
in "attention to,"
rather than living
"in a state of"
reacting to what is happening around us.

We cannot eliminate the participation
of unsuitable feelings,
unless we are monitoring our own.

Detrimental feelings surround us
and bombard us constantly.
These undesirable forces have substance
and are a threat to the "spiritual psyche."

Anything that interferes or takes away from
our happiness,
or our personal integrity,
or self respect,
is considered an "undesirable force."

We can no longer afford to allow
such negative outside influence
to participate in our embodiment,
the repercussions are too great.

Through life, if we DO allow this to continue,
the embodiment will eventually break down,
and the spiritual essence will be stifled.

Again, emphasis is placed on the "state of awareness."

When we "choose" to become
emotionally involved in a situation,
our personal decision carries,
the sense of total responsibility,
and this results in a valuable learning experience.

This art takes practice,
but once learned,
becomes a wonderful method of liberation.

To stop miscreating, know and act on feeling.

The Nature of Time Flow

All situations that we are currently involved in,
must run their course.

Situations, or life scenarios, like anything else,
have a life of their own.

Circumstances in our lives
are given the "spark of life"
through our emotional thinking.

The finished product is a "Drama,"
wholly embraced by an individual (its creator),
and allowed *by law* to exist and be experienced.

All experience is propelled by POWER,
the emotional energy inside of you.

Creating experience
from non-physical into physical reality
begins, as we have established,
with strong intentions involving
feelings and thoughts.

Many aspects must be taken into account,
to fully understand what is involved
in any existing condition of life.

These aspects begin with
the amount of emotion one gives
to a thought,
or a situation,
or a fantasy,
or a goal,
because the emotion creates the *intensity* of the drama.

Further, the emotion
can be colored negative or positive,
or have the combination of both.

Also, conditions come with an aspect of limitation.
The unlimitedness or limitedness of the condition
is determined by
our personal "beliefs."

The aspects involved are endless,
and are all contributing factors in the creation!

The duration of any drama is subject to
the continual attention
one is giving to the present condition.

Thoughts and feelings
are in a separate time flow from physical experiences.

What is in the present for our thoughts and feelings,
is in the past
for our physical body, and our physical experience.

We are literally living physically in the past,
while our immediate thoughts create our future.

Time flow exists in our world of density,
because life
PULSATES
to different speeds of vibrational frequencies.

This method of time flow is very graceful,
for if it did not exist,
all things would be created at the same time,
causing havoc in inconceivable proportions.

So, to "exit out" of an undesirable situation,
emphasis must be taken *off* the present condition,
and placed on the act of "fantasizing"
what is wanted!

Feel what you want and why you want it.

The Act of Release

The "act of releasing" pertains to,
and must be executed towards,
what one wants to create (future),
what one has already created (past),
and what one is creating (present).

It is a decision to become detatched and remain detached
from thought, feeling, and communication
toward the past and future.

Private time must be devoted
to the creation of our desires,
and then released,
for in the act of emotional releasing,
lies the power of
EXPECTATION.

This is accomplished
by envisioning a goal with feeling,
then living in the moment,
in an attitude complying with
what feels good to you.

This act of allowing our life to flow,
carries the feeling of "safety."

It controls the purity
of the new creation being formed,
as well as the static "resistance"
of the old creation that is leaving.

In this state of release,
SOURCE,
the primal energy,

is given a chance to create for us.

By executing the act of release,
we create space;
we literally create a void to be filled.

What manifests in the void,
is the next "intense" desire embraced,
whether favorable or unfavorable.

Allowing is LOVE, the premise for life.

The Power of Intent

In its purest form,
the true meaning of co-creation
is SOURCE
creating exalted experiences through us.

Feeling the safety and joy of life,
without preplanned passions,
allows us to experience this *preciousness* of SOURCE.

However, preplanned passions, along with miracles,
are our *heritage* and are needed for evolution.

Once a desired "intent" is felt within,
we need not rehearse it over and over,
unless it is sabotaged or altered
through feelings of unworth.

Disruptive events are manifested
from confusing thoughts churning through the mind,
activating unpleasant feelings.

If we recall the simple things that entered our lives,
we can observe that behind that event,
there were no worries, no cares.

What is needed from us,
is to take responsibility in the present,
live through accepting what has been created,
remove judgment to experience the good,
and recall "joyfully" what is to come.

The feelings of joy are important to embrace
because they hold the elements of
love, success, security and abundance.

The feelings of need are important to avoid
because they hold the elements of
lack, insecurity, unworthiness, and scarcity.

Intend what you want, not what you don't want.

The Laws of Love

It is an exciting journey
when we become aware of our creations.

Only when we create through the grace of Divine Source,
do we truly become masters of our destiny.

These laws mentioned and many others
pertain to the phenomenon of creation.
They allow us to be ALIVE.

They allow us to LIVE in the manner in which we do.
Everything on this planet
is a bi-product of these laws.

Our very existence was THOUGHT OF and FELT BY
something grand.

We are held together by these laws,
and we co-create with this force
in a similar manner.

We have forgotten who we are,
we have become too externalized.
We need to live life
with an "open heart" to regain our heritage.

The American Indians recognized this,
and followed these laws implicitly.

They have left us
with the only legacy life has to offer—
to live in harmony,
to recapture the ways of the earth,
and to respect her as the living entity that she is.

To live by these principal laws
is to love
the intelligence that gifted us
with the power of co-creation!

Claim your birthright.

Chapter VI

THE DECLINE OF SPIRITUAL POWER

Fear Running Rampant

Presently there is a current of fear,
a wave of fear flooding our planet.

We have been robbed by the hand of greed,
and dishonored by false promises.

For many the fear has caused their spirit to
lack energy, love, hope, and understanding.

Addiction rages for those who must "numb" their pain,
or for those who need
to "feel" something ... anything.

Violent acts have become a cry for help,
a cry to feel entitled to life.

Material comfort is now an obsession, leaving many paralyzed,
even alienated from their own intimate feelings.

This obsession has allowed some to become "robotic,"
emitting virtually no life force
from their "being" whatsoever.

Then we have a class of people
who have very little meaning for life,
who eliminate life with little to no concern at all.

These classes of dysfunctionality
ripple out and effect everyone on this planet.

Perform an act of kindness every day.

False Power

Deep within humanity
there is a suppression of knowledge
which has been allowed to persist for centuries.

This suppression that has handicapped our spirit,
has *turned in on itself*
causing turbulence of both
large and small proportions.

All thoughts and actions have a substance,
therefore are *felt* and *transmitted* from us.

The life span of a negative attitude, along with
the combination of anger, hatred and greed,
have made a direct impact
on the inhabitants of this planet
leaving them weak and fragile.

Our gullibility has allowed the leaders of society
to perpetrate the idea
that "all is well,"
when *all could be much better*.

A system that supports natural foods, rather than
processed foods, *contributes* to life.
A system that designs aid programs to "motivate,"
rather than handicap, *contributes* to life.
A system the focuses on truth, rather than
sensationalism, *contributes* to life.

Our system of living intervenes rather than prevents.
Our world is addicted to and manipulated by control.

The most powerful means of control

are criminal, religious and political,
but there are many more.

We need a new VISION for our world.
We need a new way of understanting LIFE.
We need to VALUE life.

All we need to do is
take an interest in "what's going on"
with our precious animals, fellow human beings,
and our Mother, the Earth.

The discovery of our "own power" is greatly needed.
To live in ignorance and control is destructive.

False power is a force
that does not sustain life,
but consumes it.
We have not come here
for what we "think" is a short period of time
to "take" then "leave,"
It's much *deeper* than that.

We need a new perspective of the world.
We need to understand that
we are connected to our planet
not separate from her.

If we injure her we injure ourselves.

The time has come for us to use our wisdom.

Life Weaving Life

There is a biological, psychological network
of activity going on faster than realized
between us and every living organism
on our planet.

On a deeper level still,
our connectedness is *subatomic* in nature,
seeded genetically through *emotional principles*.

Our manipulative ways have created
a burden on others that need to survive,
as well as on future generations.

The wisdom of the native American Indian
was an extraordinary perspective of life indeed!
It was a life style "aligned" with nature.

They clearly demonstrated this alignment,
by making decisions based on
the "thriving" of life
as opposed to the "misuse" of life.

It is documented that a great interest and reponsibility
was taken in preparing for future generations
by passing down their teachings and tradions
to their children.

Hopefully we will allow their wisdom
to return and penetrate our lives today.

The responsibility is always
in the hands of the individual.

Respect is an exchange of love

for *all* that nurtures us,
human to human
...human to animal
...human to earth.

When will we learn and educate ourselves
on the mechanics of the earth?
The game of ignorance has been played out
too long.

Love life, we can no longer afford to reduce it.

The Nature of Things

If evolution came prepared with its own script,
Mother Nature
would have the leading role and steal the show.

The natural disasters occurring on our planet
are an ingenious way for earth to self maintain,
and bring humanity closer at the same time.

There is nothing more profound
than coming face to face with death
to cause change within us.

For some "survival reason,"
the importance of life
makes itself known in an instance.

In the midst of a life threatening situation,
great power and awareness is unleashed,
and LIFE becomes *the primary goal.*

Unfortunately, sometimes it takes a brush with death,
to realize the importance of living and helping others.

When one *unconditionally* helps another,
the LIFE FORCE SYNERGIZES within both,
and great things are created
in places unseen to the naked eye.

This is the "intensity"
that allows us to feel life within us.

To gaze with love into the eyes of another human face,
is to capture the GOD within the Wo-Man!
There lies the power,

in that knowing, in that respect,
in that EQUALITY for one another.

Nature seems to be taking care of her part;
now we need to combine mental awareness
with physical environmental changes,
to bring about restoration on our planet.

Deep inside, when one gifts themselves with
moments of solitude and peace,
one can feel the safety of life,
and the fear of non-existence is removed.

This feeling of safety is
a profound knowing,
a promise.

Each and every one of us is truly loved,
we must begin to remember and share that love.

The power of Love is perpetual.

Hidden Forces

There are forces all around us, manipulating humanity.

Many of these forces have succeeded,
  and have flourished for a very long time.

They come in many guises, visible and unvisible.
They are kept alive and *empowered*
  by our **fear** and ignorance of change.

Ignorance has been the fuel for our current state of affairs,
  however,
  it CAN be eliminated very easily.

Ignorance is empty,
  recycled space waiting to be used.
Ignorance is a lack,
  an ability performing as an inability.

What has been performed by the "puppet,"
  is whatever the puppeteer wants.

However, *aspiring energies* living within us
  are trying to surface.
These energies that live within us
  come from a space of purity.

Although repressed from expression in this world,
  these energies have been allowed to *create*
  on many different levels of existence
  not visible to the human eye.

These levels are valid in their existence,
  and carry an unparalleled band of energy
  unfamiliar to the forces

that have been allowed to reign
on this plane of existence.

These aspiring energies will aid
in promoting *harmonious change* on our planet.

Energy cannot be created nor can it be destroyed,
so the question is, "What will happen
to the diverse energies?"

The answer might be,
that these forces will go into remission
until such time that they choose to evolve.

One thing is certain,
the virtue of Love is everlasting,
and much needed on our planet.

Release the Puppeteer and become the Master.

Chapter VII

LIBERATING THE FIRE WITHIN

I Am

The Power within, is a *feeling* of Exaltation.

It is much deeper than the love
we experience and call "human" love.
It is derived from the Soul,
and can be labeled a Personalized Endorsement.

It is an "inscription" of energy
allowed to *participate* in human form
to *gain* and *feel* "itself."

On our planet, **IT** adventures through the personality.

IT filters through the mentality,
and is experienced through the physical senses.

IT is *empowered* by feeling,
creating our *tool* the EGO,
by its ability to recycle itSELF
into a personality.

When our ego feels "safe,"
it is embarking on the journey it was intended to take.

This feeling that is Power,
this feeling that is Love,
is an "essence" to be discovered
and externalized
in our world by each and everyone of us.

It is to bloom and pollinate like our flowers.
it is to procreate amongst us
for the creation of an enlightened species.

This is who we truly are.

Revitalize and transform the earth.

The Life Force

This business of looking within
can be such a foreign request for so many,
yet our very existence as a being
depends upon this "Feeling."

This power exists on a scale
inconceivable to the human mind.

Embracing that information alone,
would give us many answers,
however, we need to start in the stages of TRUST,
for the "puppeteer" has held the strings too long.

We shall call this "Feeling" THE FIRE,
for when knowledge of its awareness is felt,
it ignites within us
and an emotional sensitivity
penetrates every cell of the body.

It carries within itself the greatest
INTENSITY OF LOVE
ever imagined or felt.

This intensity has levels,
and can *only* be experienced through the *feeling tone.*

It expresses itself through our personal "signature,"
creating unique qualities in personalities,
while encompassing life with great variety!

It is power in its purest form,
precise in nature,
focused and poignant in direction,
absolute and flawless in its performance of knowing,

complete in its presentation of energy,
and virtually unlimited
in its range of probability.

This power lives in you, the Master.

You The Director

Accepting and giving *reverence* to
the Fire living within
is of utmost importance.

Without this acknowledgement,
there will be no basis to begin
and its presence will remain dormant.

We are always the director in our life,
that is inescapable.
We are always creating a pathway,
always in a state of decree,
regardless of what we are decreeing (or surrendering).

That is the role of this energy,
and if it wants to shut itself *off* from itself,
it can do that also!

One might wonder "why" it would do that,
but in truth, it only appears
to be shutting itself off,
because its awareness is absent.

In essence, it is *always* expressing
regardless of the personality
living redundantly
(spinning our wheels).

It stays in a continuum
until it desires more or something different:
this is why awareness is so important.

Awareness creates movement while eliminating pain.

This energy permits
total life expression to occur in everything we do.

The process of discovering and developing
this Fire within,
depends on the very willingness to desire it.

This willingness determines or "measures"
the amount of Fire
we are willing to *feel* or *hold*.

Everyone's level of intensity
will measure according to its use and practice.

The act of living with this Love flowing through,
requires courage:
Courage to confront and break free
from all the "puppeteers" in our life.

It is always flowing, claim it.

The Fire of Love

The "Fire" is a thrust of energy that
breathes in, through, and out of you.

The degree of Fire that can be held individually,
depends on "conscious awareness" at work.

Keeping it simple is always the best method.

Ponder on the fact that
we were created from something wondrous,
and this "something" loves us.

Turn that belief into a *knowing*, a pathway
by asking internal questions
and listening for internal answers.

Learn to feel that this *knowing* is taking care of you,
nurturing you,
as you become relaxed in everything you do.

This Fire of Love is
very knowledgeable, very experienced.
It has created everything here, and in the universe.

When we trust it,
through the ACT and APPLICATION of courage,
great insight will be revealed to us.

The Fire will transform and change you forever.

A Walk Through Nature

Nature is a wonderful place
to begin feeling this energy.

Take the time now to be still
and meditate on the following words.

Visualize and "feel" one segment at a time,
as you allow yourself to play,
in the field of your imagination.

Picture yourself taking a walk ...
walk for as long as you like.
Now sit and rest for awhile ...
while sitting observe your surroundings,
for there are many things to be seen in nature.
Breathing in the sweetness of the air,
the comfort of the day has you very relaxed ...
now find yourself
succumbing to the Trees that shade you ...
succumb to the Plants that surround you ...
succumb to the Rocks that support you ...
succumb to the Animals that call to you ...
succumb to the Lake that will cool you ...
succumb to the Mountain that inspires you ...
succumb to the Clouds that dream with you ...
succumb to the Sun that warms you ...
succumb to the Moon that mesmerizes you ...
and succumb to the Stars that whisper to you ...

Gift yourself with this exercise
as often as possible, however,
it is more effective to place one's self outdoors.

Take the time to
practice feeling
the "reflection" of nature within you.

Allow the memory of these feelings
to resurface frequently,
for not only do we feed ourselves in this adoration,
but we feed Mother Earth as well.

Embrace all of nature's essence.

Waking Up To Yourself

By believing and adopting the concept
that *everything operates as a reflection*,
as a mirror to everything else,
we can begin learning that
the universe adapts itself to our view of reality.

Sometimes what appears to be abhorrent in our lives
can be a great catalyst
in our search for true identity.

Under many layers of "puppeteer" conditioning
lies a spark,
a spark as powerful as a Super Nova!

The Fire within must be felt **consciously**,
and to feel it consciously
requires acquiring great "intensity."

This intensity is formed through the process of
understanding and *loving* our self.

A wonderful tool for this "awakening process" is
the glass mirror!

The following exercise is to be practiced
during a quiet time
and without interruptions
for its full benefits to be received.

This is an experience in purging,
purging the layers that we hide behind.

The amount of time needed is irrelevant,
for it will be the amount of honesty taken

that will serve us best.

Let's begin by gazing into a mirror.
Using passion, allow the desire to "know who you are"
surface *within your body*.
Give thought to who you "think" you are,
what you have become for others.
Now give thought as to what you have become,
for yourself.
If you come up short for yourself,
create the passion to know what that would be.

Still gazing begin to speak aloud these thoughts.
Experiences of deep aloneness may arise,
allow them.
Feel free to feel what's in your heart,
for the "Fire" within will comfort you.
Raise the intensity,
allow yourself to feel and voice
any fears, anger, or desires you might have.
These emotions contain
judgements, misunderstandings,
disappointments, betrayals, lies,
unfulfilled hopes and wishes,
held within, caused by others and yourself.
Release these emotions,
they are poison to the body.
Dissect them one at a time.

If exhaustion sets in,
stop and continue at another time.
Begin to calm yourself by breathing slowly.
Still gazing into the mirror,
address yourself by name;
it denotes a gesture of love and respect for your self.
Take time to gaze into your eyes (profoundly).

While gazing honor yourself with love,
allow the love to pour from your eyes
to your image in the mirror.

Ask yourself the question, "What do I want in my life?"
Wait for an answer.
Ask yourself what will make you happy.
Wait for an answer.
Ask yourself what you are willing to change.
Wait for an answer.
As the answers come,
penetrate deep into your eyes,
and surrender to your heart.

Allow the feelings of suppressed desires to stir.
Allow the dreams not permitted
to be fantasized, to be felt.

Let the tears flow,
hold on and never stop the gazing.
Look for the love,
and don't stop until you find it ...
Soon all the seriousness will subside,
and a beautiful smile will appear on your face.
reminding you of your preciousness.

As you continue to gaze into the mirror,
the comfort will return,
and you'll feel stronger, nurtured, lifted.
The tears of pain
will have transformed into tears of joy,
as you stand WITNESS to your POWER!

This is the Love of God,
the Love of Source,
reminding us that we are PERFECT,

that we are LOVED, *forever*.

Listen to the Fire and become the magician.

Evolution At Work

The Fire within bestows upon us
creations of life.

These new experiences can only be obtained
through our ability to *receive* love.

Do not fall into the trap of understanding
this energy ONLY as a concept,
for nothing will change,
nothing will occur,
unless it is FELT.

The use of the mental faculties
is to provide a point of reference
to concepts ALREADY experienced.
A concept already experienced
is a "frozen" block of information,
derived from emotional understanding and feeling.

Mental concepts reside and are "pulled" from
your personal library
called the Subconscious Mind.

Personal growth can, however,
bring on a slightly different slant to a concept,
but in respect to the Fire within,
the concept is "lodged" in its own particular meaning,
and therefore is considered OLD.

A concept already experienced,
lives in the PAST;
the FIRE within lives in the PRESENT,
while creating the FUTURE.

Mental concepts provide
the memory of our position for reevaluation.

Utilizing the Fire
creates the next move for NEW action.
When working with this energy,
the next "feeling tone" is authentic,
and therefore can not be *labeled*.

When new information is created upon,
an "implosion" occurs
within the body through the nervous system.

This new creation then becomes an experience.
Once an experience,
it receives a label by the mind
and is then filed in the subconscious.

To deny our self the expression of this energy
is a spiritual tragedy,
a "drama" we can no longer afford to have.

Love yourself into LIFE.

Chapter VIII

ACKNOWLEDGING GUIDANCE WITHIN AND AROUND US

Demons of The Past

To LIVE life, is to be in a constant state of
feeling oneself living.

Behavioral patterns stop the flow of life;
they are deadly and attack the "pulse of life."
These patterns of "inflicted" behavior
have been "learned,"
and are only habits to be removed.

If we are living with any feelings of pain ... or confusion,
then we are in "denial"
of this LOVE, of this POWER, of this FIRE.

To blame is an inappropriate act,
for the one being blamed has also learned
to become a victim of his past learning.

Guilt is also an inappropriate act,
for the guilty one has also learned
to become victim to her past learning.

Emotional pain derived from any experience;
if held too long will run deep.
It can act like "hypnosis"
clouding and "possessing" our perception
and creating all sorts of conflicts in life.

As long as there is "denial" of this LOVE,
the behavioral patterns will move in a cycle.

Pain takes away life, Love always gives life.

Finding Our Essence

It is time to start responding for our self
in ALL areas of life.

This sense of security comes from within,
and as a reward brings
many things that are desired and loved.

Achievements and composure levels will increase,
feelings of well being will be experienced.
Rest will take the place of struggle,
and composure will replace resistance.

We must "pull into" our self:
Attention must always be placed upon the *essence*.

We need not look to others
to find worth within ourselves.
Drawing our attention inward
creates a *natural centering* within the body.

This "act" of centering always places us
in the conditions suitable for
growth, joy and creativity.

Trust every feeling and every word "revealed"
in this space of centeredness.

Recognize what is wanted (by you),
rather than what others want from you.

Respect and respond to others with love,
as you TRUST YOURSELF.

Have the courage to respond and "voice"

all feelings with truth and love.

Have the courage to "act" on all feelings,
if action is needed.

Remember,
happiness exists in the *action of love*:
an existence of LOVE,
for our self.

This is the only power!

Experience the thrust of life!

The Guidance of Earth

When we are in a state of centeredness,
or simply sensing our well being,
we are connected to GUIDANCE.

Guidance can be experienced in many forms,
seen and unseen.

The guidance of our earth is one we can truly see;
it reveals itself through the elements,
Earth, Fire, Air and Water.

The element of the Earth
reminds us that *we* and *everything*
in nature is continually *self-renewing*.

Observe the seasons:
Spring is the beginning of new life.
In Summer, that new life "peaks" in creativity.
After reaching its full expression,
Autumn arrives and we "reap"
the benefits of that creativity.
As the Dance of Life calms down,
giving sway to Winter, life turns "inward"
to sleep and *dream* of its new creation once again.

This continuous story of nature
reveals and models to us the format of life,
and serves as great guidance.

As we observe spring,
we are reminded of the sower,
for our thoughts do act as
powerful "seeds" upon our land.

Summer begins and our seeds start their growth.
After culmination,
we experience new life,
new activity from *our* creations.

Autumn appears and the celebration begins
as we marvel with delight,
embracing the fruits of our labor.

Winter sneaks in and sends us to rest
to contemplate again new seeds for thought
while dreaming our next dream.

The element of Fire
encourages us to live life **passionately**
and with fervor.
Nothing on our beautiful planet
is created without passion,
for passion is the **thrust** of life.
The more intense the passion,
the greater the manifestation.

The miracle of *giving life*
is the most passionate creation bestowed upon man.
With all its intricacies and probabilities,
new life is chosen and emerges through us.

The Fire within is this continued passion
burning secretly inside
urging us to raise its flame brighter
guiding us toward new horizons.

The element of Air
appeals to the mind,
to our sense of wonder,
and seeks out our wisdom.

Its wayward breezes carry the gifts of
Intuition, Inspiration, and Dreams.
Its mysterious breath echoes antiquity,
and reveals *all knowledge*.
In its sweetness
lie the hopes and desires of the future,
for the air guides us
to rendezvous with the Soul
spiraling toward expansion forevermore.

The element of Water
unravels our emotions in its timeless flow,
and returns them back to us
full of compassion and understanding.
The roaring seas,
the quiet lakes,
the winding rivers,
all carry tales of love and sorrow.
One thing is certain,
its message runs deep,
for its essence is **feeling**.
Feel the water's power,
for it holds the map
leading to the *greatest treasure* ever found—
the treasure of the *heart*.

Therein lies the grandest story ever told,
YOU,
in all its bounty.

Nature's voice is always speaking.

The Guidance of Spirit

The guidance of spirit
comes to us in many forms and profiles,
whether religious or mystical symbology,
nymphs, angels or visions.
We are *never* alone or without help.

Guidance provides wisdom,
and this wisdom is felt in the heart.

Guidance that is not in alignment with the feelings,
needs discretion.

There is a very important factor:
an "openness" to receive guidance must be present.
Without this desire or intent,
guidance will not be "received."

This method is very simple
and remains the same for everything wanted.
Simply desire guidance,
then "shift" the attention
to a feeling of "openness."

The time flow in receiving guidance varies,
it can happen anywhere from a moment to
a period of weeks.

Guidance presents itself in various ways.
It may appear through dreams,
other people,
through subjects of interest,
in nearby or far off places.
The variables are endless, but the guidance will come.

When guidance appears externally or "out of the blue,"
trust and recognize this information
to be accurate,
for it was "sourced" by you.

We may choose not to use it,
but be aware that it holds "a key"
toward the direction we are desiring to experience.

Not using the guidance sent to us
will only hinder us from
attaining our goals.

Our goals will be put on "hold" so to speak
because acting on guidance produces movement
"toward" the goal.

When intentionally seeking outside guidance,
use discretion by sifting through
all the information received.

Capture what "feels" right to you
discard the rest,
and make the information your own.

There is always a slight risk
when listening to others for guidance,
for one can be influenced in a different direction.

We must be careful of what we *embrace*.
Being influenced by outside guidance
can temporarily sidetrack us from our goals.
However, outside guidance
is suitable to confirm our decisions.

Contemplate the guidance from Earth,

and the guidance from Spirit,
for they provide us with many things for many reasons.

The surest way to benefit from them and all things,
is to feel their appreciation.

Guidance serves us well,
when we are in a state of appreciation.

The unfailing source of guidance is always your own.

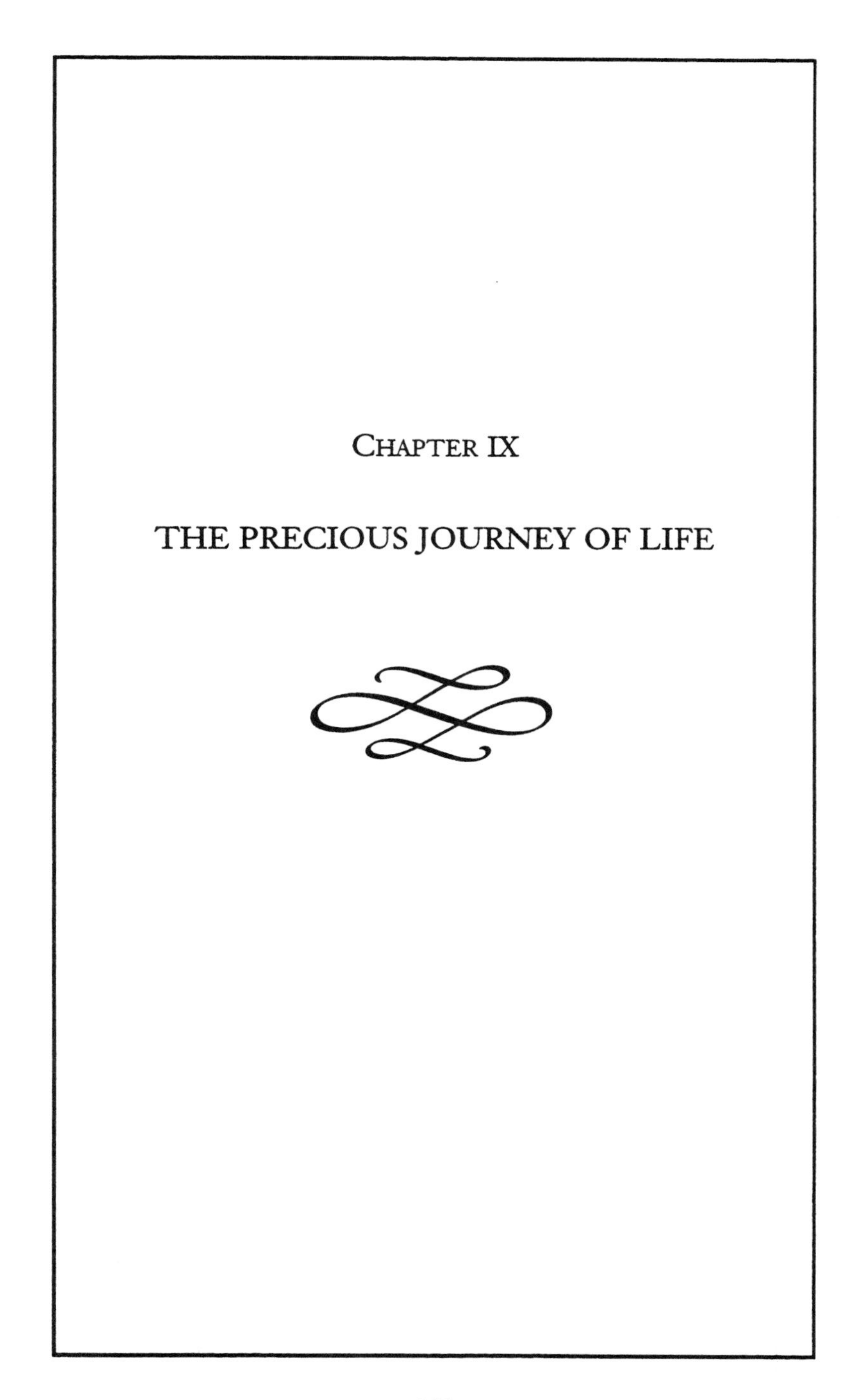

Chapter IX

THE PRECIOUS JOURNEY OF LIFE

An Adventure Worth Taking

To inhabit a human body
and fulfill its function
—the gift of physical experience—
is a wondrous thing indeed!

We are here to journey into the experience called LIFE.

We are here to value life,
and we are here to CREATE life.

There is only one test, one lesson set before us.
Yet it is not a lesson, but a "model,"
and not a test, but an "investigation."

Without it one does not fail,
one collapses,
for it is the originator Love,
and we are its offspring.

Our role as **creators**
is to expand the intensity of love on our planet.
But first we must *discover*
our selves through that love.

Life is a creative process
that requires *devotion* to our self.
The value and support we give to our self
for living our life,
is the very essence of life
and creation.

There is only one thing that contributes
to the feeling of "powerlessness," and that is,
not owning the knowledge of who we are.

Restoring power can only occur
through THE EXPANSION AND CAPACITY
TO RECEIVE LOVE.

Knowingness is a requirement
and derived from the "act" of
respecting our knowledge into existence.

Living in alignment with the planet is an "*inside job.*"

The attempt and practice
of personal integrity with love,
reaches out, overlaps,
and like anything else, spreads.

We then can experience life together,
and while maintaining our individuality,
we can become true co-creators with our planet.

Begin the journey by
living in the feeling of **Beauty**,
living in the feeling of **Laughter**,
living in the feeling of **Enthusiasm**,
living in the feeling of **Exploration**,
living in the feeling of **Gratitude**,
living in the feeling of **Joy**.

When feeling lonely,
GIVE MORE LOVE.
When feeling hopeless,
GIVE MORE LOVE.
When feeling sad,
GIVE MORE LOVE.

We are to **value** ourselves
and accept others;

Author's Note

We live at a time when life is very active yet incredibly excessive. We live at a time when we need reminding of the priorities that will sustain us, that will nurture us, that will support life. We live at a time when we must recognize that we are indeed more than a body of flesh— we are a body of spirit as well, and our lives are about the unfolding of this spirituality. We live at a time when life can be very exciting, very miraculous, very powerful for those who decide to engage and live within their spirituality.

There was a time on our land when a rich spirituality existed. It was the time of our brothers and sisters, the native American people. Throughout the book I make reference to the native people, for to me their spirituality contains a profound "purity" of knowledge and a great "intensity" for the value of life. They are held in mind and spirit as a great source in the creating of our own deep spirituality here and now.

It seems our essence unfolds when we direct love and respect toward ourselves and others. *A Voice For The Planet, A Voice For Humanity* was written to motivate and create a sense of courage for those who dare to leap into the discovery of themselves. Each one of us must find our own purity of knowledge, our own intensity for life. It is not out there in the world of form; the mystery lies within. Discover the mystery then pass it on!

to a "living system" as a whole.

As the journey continued,
new systems were formed,
and the creatures exceeded
into the infinite in their count.

One fine day, glorious beings emerged,
traversing through this garden of life.
As these beings observed
the activity around them and with each other,
they were filled with wonder and delight.

They had soon learned
to interact with their surroundings,
and soon discovered that they too,
had a creative role in this world of abundance.

From the participation of the gifts available to them,
they had come to an understanding
that they were "partaking" from the system,
and they recognized a need for balance.

They too possessed great intelligence
and instinctively realized
that to sustain the system,
they needed to develop a system of "exchange."

Through this process they witnessed
a *fostering connection* between themselves
and their providers.

A new sensation arose within them,
and they felt a tremendous
sense of well being and contentment.

They began to grow in number,
for as their harmony increased, so did they.

Much to their surprise,
their variety blossomed both in color and shape
adding more enthusiasm to their existence.

Frolicking through time and space,
their inventions grew
and wondrous creations were experienced.

Maintaining their respect for themselves
and for their providers of the global garden,
they began to expand in their perception.

At that time,
a new gift had been bestowed upon them
from their creator force:
LOVE, the gift of self-realization.

They had come to "realize"
that they themselves possessed
the very force that birthed them and all things
into existence.

For until that moment, they had lived
as a "vision within the dream" itself,
unaware of the power that housed them.

Once again,
they were filled with feelings
of exaltation and fascination.

Once again, their creations began to take on
new shapes and forms,
for they had discovered "experience of vision"

allowing them to construct their "own" creations.

Growing in passion from their new power,
something curious happened to everything
around them.
They began to observe an energy field
surrounding whatever they were "beholding."

This was very exciting to them,
for the energy field
presented itself in various shades of color.

They noticed that, not only was the energy surrounding
what had been created,
but it was also surrounding the beings themselves.

The energy was vibrant and changeable.

They knew this phenomenon
would add scope to their visions,
and within that embrace, that moment,
they were gifted once again with a new insight,
Comprehension of Feeling.

As they practiced "feelings,"
they once again opened up to new creations.

They were now producing inner worlds
as well as outer worlds,
knowing that their possibilities were endless.

Well, they came to a great understanding one fine day:
that the realms of creation are sustained
by a design inherent in the process of love,
and that by maintaining respect for that love
they were able to become creators themselves.

With this revelation,
they continued their journey
into realms of exquisite vision,
while knowing that their essence
known as love,
would continue into journeys of exquisite vision ...
FOREVER.

Remember.

Into Eternity

The dance of life is simply,
—Love your visions, visions create movement:
Movement occurs through the act of non-resistance.

—Trust your feelings:
Feeling is the fuel that ignites movement.

—Choose what influences you:
Choice is the force that sustains momentum.

These words have been written to move us
into decisions of change,
for peace within our selves,
peace with our loved ones,
and peace within the world.

Every thought, act, and feeling
of respect given to others
and most importantly to ourselves,
acts as seeds spread upon our land.

Spread these seeds
and keep our planet healed once and for all.

The Dreamer and the Dream become one.

My Message to You

There will be many questions along the way;
some will have answers and some will not.

The ones that do not must be handled with courage,
but do not discourage,
the answers will come in their time to encourage.

There will be many ways offered
and some you will choose,
but if the processing tampers at all with your views,
allowing your bliss to become confused,
put them aside, they are not to be used.

Remember, the rules that apply
are the rules of the heart.
This my dear friends will give you a "great start."

This is the next step of the journey you see,
the next step into life is to BE!
Value it deeply, choose wisely my friends,
for happiness is our promised birthright
that will *never* end.

Remember the dreamer, remember the dream.
Remember that you are loved,
and forever you'll be in the **main stream**.

Go play.